THE WORLD OF OCEAN ANIMALS

JELLYFISH

by Bizzy Harris

pogo

Ideas for Parents and Teachers

Pogo Books let children practice reading informational text while introducing them to nonfiction features such as headings, labels, sidebars, maps, and diagrams, as well as a table of contents, glossary, and index.

Carefully leveled text with a strong photo match offers early fluent readers the support they need to succeed.

Before Reading

- "Walk" through the book and point out the various nonfiction features. Ask the student what purpose each feature serves.
- Look at the glossary together. Read and discuss the words.

Read the Book

- Have the child read the book independently.
- Invite him or her to list questions that arise from reading.

After Reading

- Discuss the child's questions. Talk about how he or she might find answers to those questions.
- Prompt the child to think more. Ask: What did you know about jellyfish before reading this book? What more would you like to learn?

Pogo Books are published by Jump!
5357 Penn Avenue South
Minneapolis, MN 55419
www.jumplibrary.com

Library of Congress Cataloging-in-Publication Data

Names: Harris, Bizzy, author.
Title: Jellyfish / by Bizzy Harris.
Description: Minneapolis: Jump!, Inc., 2022.
Series: The world of ocean animals
Includes index. | Audience: Ages 7-10
Identifiers: LCCN 2021022619 (print)
LCCN 2021022620 (ebook)
ISBN 9781636902821 (hardcover)
ISBN 9781636902838 (paperback)
ISBN 9781636902845 (ebook)
Subjects: LCSH: Jellyfishes–Juvenile literature.
Classification: LCC QL377.S4 H37 2022 (print)
LCC QL377.S4 (ebook) | DDC 593.5/3–dc23
LC record available at https://lccn.loc.gov/2021022619
LC ebook record available at https://lccn.loc.gov/2021022620

Editor: Jenna Gleisner
Designer: Michelle Sonnek

Photo Credits: Eric Isselee/Shutterstock, cover; Antonio Martin/Shutterstock, 1; Richard Whitcombe/Shutterstock, 3; rebius/iStock, 4; Beautiful landscape/Shutterstock, 5; Damsea/Shutterstock, 6-7; Minden Pictures/SuperStock, 8-9; efilippou/iStock, 10-11; Shutterstock, 12; borzywoj/iStock, 13; WaterFrame/Alamy, 14-15; Martin Prochazkacz/Shutterstock, 16; cbimages/Alamy, 17; Felix Agel/SuperStock, 18-19t; Jordi Chias/NaturePL/Science Source, 18-19b; Ethan Daniels/Shutterstock, 20-21; Arina_Bogachyova/iStock, 23.

Printed in the United States of America at Corporate Graphics in North Mankato, Minnesota.

TABLE OF CONTENTS

CHAPTER 1

BEAUTIFUL BELLS

A jellyfish drifts in the ocean. Its **tentacles** hang down.

Muscles around the jellyfish's bell tighten. Water is pulled into the bell. Then the muscles relax. Water is pushed out. This moves the jellyfish forward. This is how it swims.

Jellyfish come in many colors and sizes. Some live in warm, shallow waters. Others live in the deep, cold sea.

TAKE A LOOK!

Jellyfish live in every ocean. Take a look!

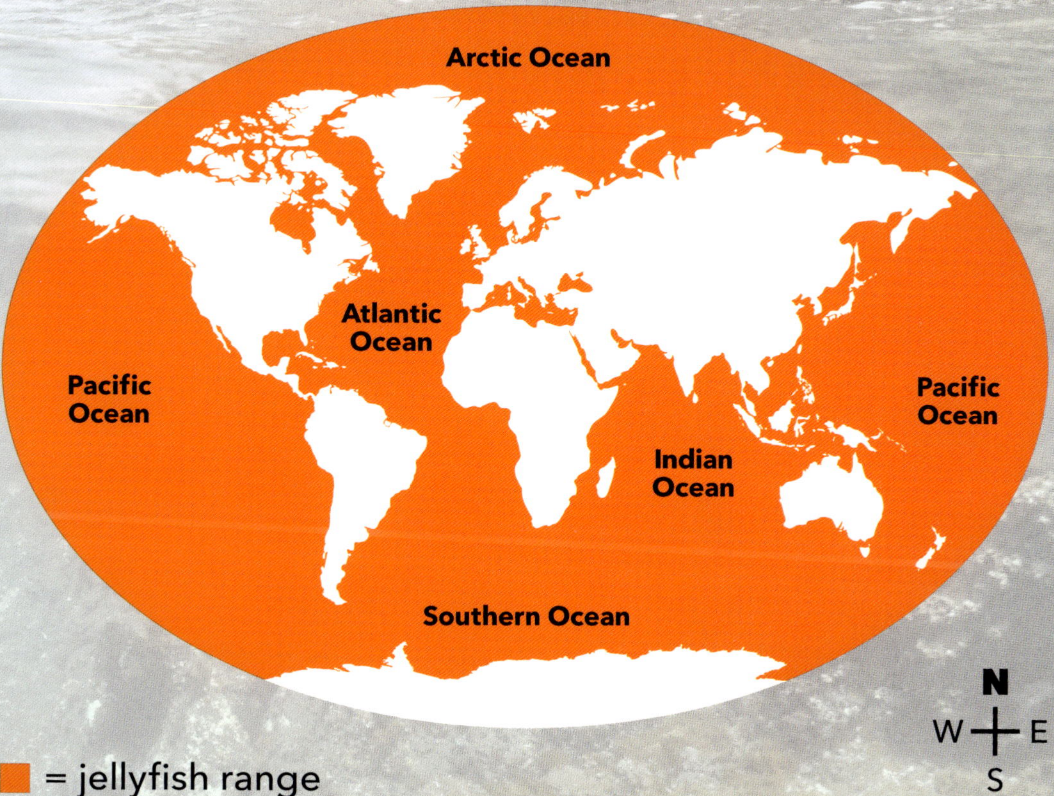

■ = jellyfish range

A jellyfish starts as an egg. Then it becomes a **polyp**. It attaches to a hard surface. Parts of it break off. These are young jellyfish. They grow into adults and are called medusae. Stinging tentacles and **oral** arms help them catch food and eat.

TAKE A LOOK!

What are the parts of an adult jellyfish? Take a look!

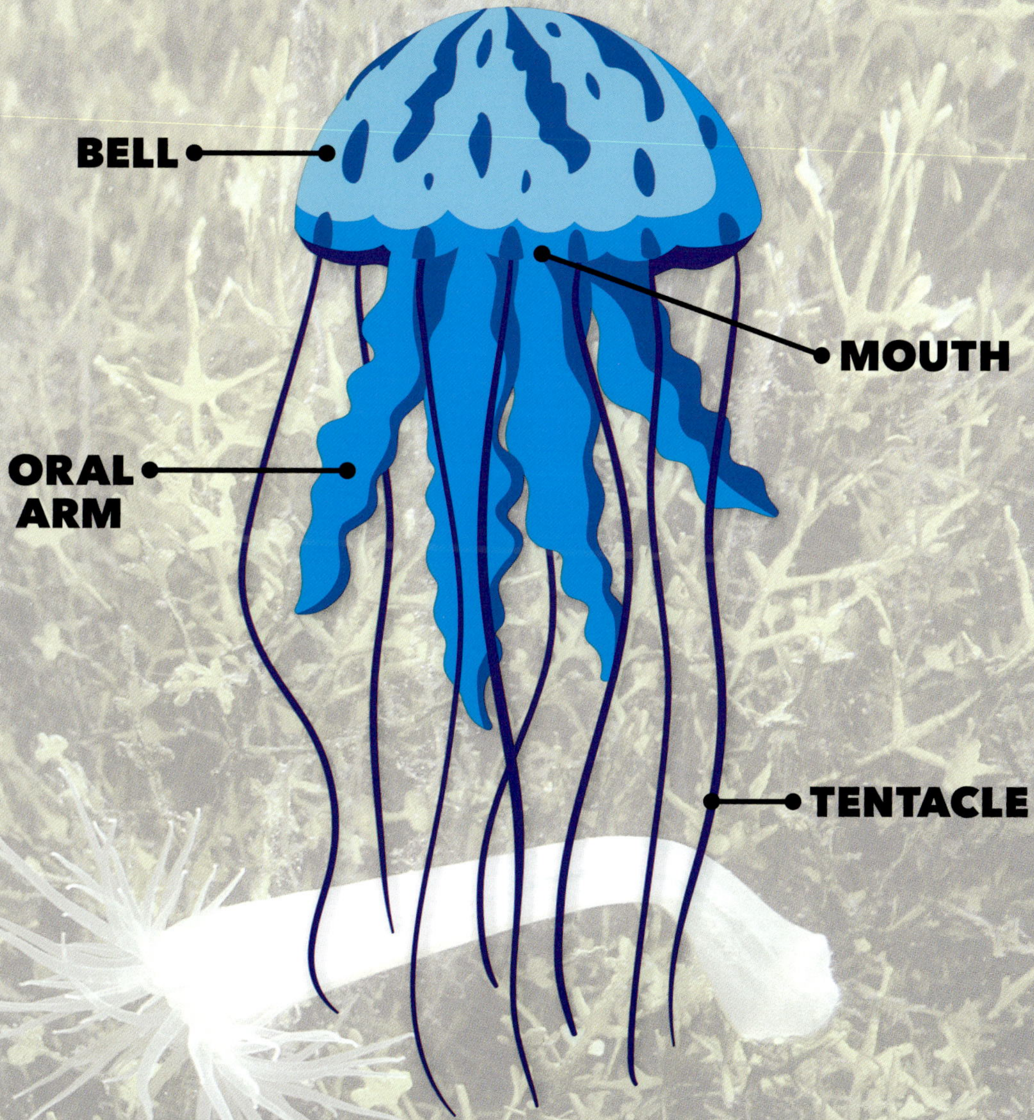

Despite their name, jellyfish are not fish. They are **invertebrates**. They don't have any bones. In fact, they are about 95 percent water!

Their skin is thin. Through it, they **absorb** oxygen from the water.

DID YOU KNOW?

Jellyfish don't have brains, hearts, or eyes. Instead, they have a nerve net. This is a kind of **nervous system**. It allows them to **sense** things, such as light and smells.

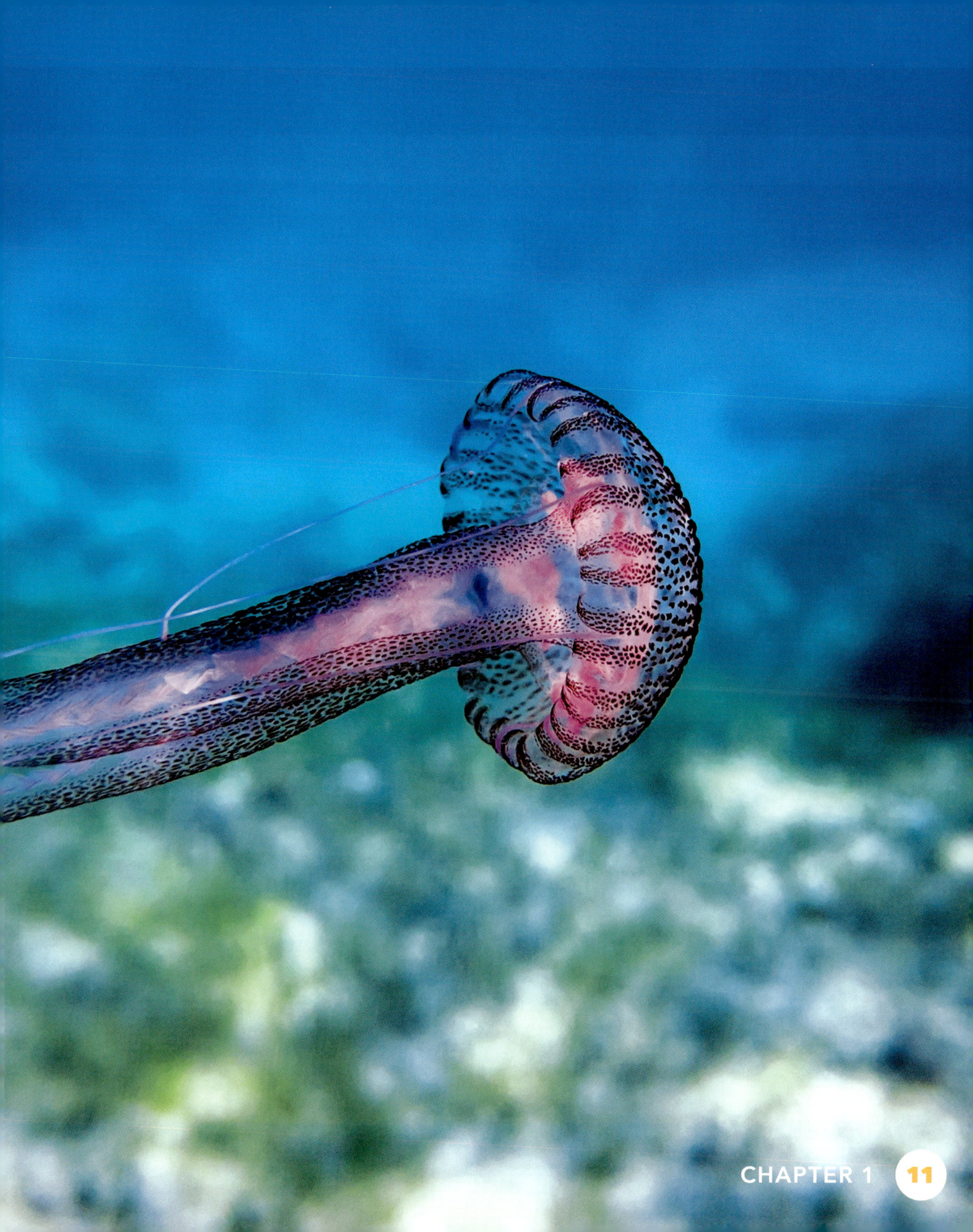

CHAPTER 2

POISON, PREDATORS, AND PREY

Most jellyfish are see-through. They blend in with their surroundings. This helps them hide from **predators**, such as sea turtles and fish.

Jellyfish are **carnivores**. They mainly eat **plankton**. Larger jellyfish sometimes eat fish, shrimp, and other jellyfish!

Jellyfish tentacles shoot poison. The poison **paralyzes** their **prey**. Their arms help move the food toward the mouth. Once food is **digested**, this is also where waste comes out!

DID YOU KNOW?

Jellyfish can lose their tentacles. Sometimes, they wash up on the beach. Don't ever touch or step on one! Why? Lost tentacles can still sting!

CHAPTER 3

KINDS OF JELLYFISH

There are around 200 jellyfish **species.** The largest is the lion's mane jellyfish. Its bell can be eight feet (2.4 meters) across. Its tentacles can be 120 feet (37 m) long!

One of the smallest is the thimble jellyfish. It is named for a thimble you put on your thumb! It is no bigger than one inch (2.5 centimeters) across.

Some jellyfish light up! The moon jellyfish is one. The Atolla jellyfish is another. It flashes when it is attacked. Why? Scientists think it is trying to confuse predators.

moon jellyfish

Atolla jellyfish

bloom

Groups of jellyfish are called blooms. These are becoming more common. Why? **Climate change** could be one cause. Warmer oceans mean jellyfish can grow quickly. They can survive in **polluted** waters. Would you like to see a jellyfish?

ACTIVITIES & TOOLS

TRY THIS!

BLENDING IN

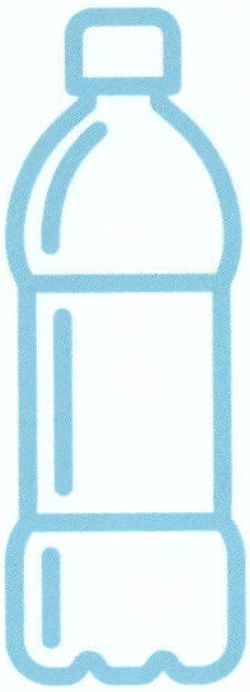

See how a jellyfish blends in with its surroundings in this fun activity!

What You Need:

- clear plastic bag
- scissors
- clear plastic water bottle
- piece of string
- blue food coloring

1. **Place the plastic bag flat on a hard surface. Use scissors to cut off the handles.**
2. **Cut the bag in half so there are two separate sheets. You will only use one sheet to make your jellyfish.**
3. **Bunch up the plastic to make a balloon. This will be the jellyfish's bell. Loosely tie a piece of string around the base of the balloon. Leave some of the plastic sheet hanging down from the balloon.**
4. **Below the string, cut the sheet into strips. These will be the tentacles. Make as many as you'd like!**
5. **Fill the balloon with water. Leave a little bit of air inside. This will allow your jellyfish to float! Then, tie the string tight.**
6. **Fill your water bottle. Place your jellyfish headfirst into the bottle. Add a few drops of food coloring. Secure the lid and shake. Observe your jellyfish. Does it blend in with its surroundings? Why do you think that is?**

GLOSSARY

absorb: To take in.

carnivores: Animals that eat meat.

climate change: Changes in Earth's weather and climate over time.

digested: Broke down food so that it could be absorbed and used by the body.

invertebrates: Animals that do not have backbones.

nervous system: The system of nerves in the body that sends messages for controlling movement and feeling.

oral: Of or having to do with the mouth.

paralyzes: Makes someone or something unable to function.

plankton: Tiny animals and plants that drift or float in oceans and lakes.

polluted: Contaminated or made dirty or impure, especially with products made by humans.

polyp: A small sea animal with a tubular body and a round mouth surrounded by tentacles.

predators: Animals that hunt other animals for food.

prey: Animals that are hunted by other animals for food.

sense: To feel or become aware of something.

species: One of the groups into which similar animals and plants are divided.

tentacles: The flexible limbs of some ocean animals.

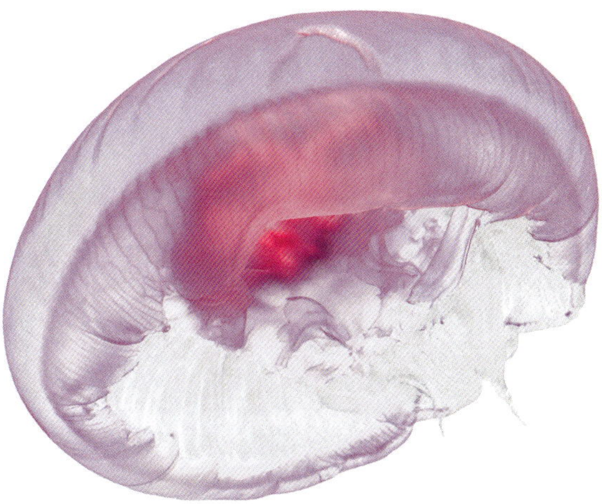

INDEX

TO LEARN MORE

Finding more information is as easy as 1, 2, 3.

1. Go to www.factsurfer.com
2. Enter "jellyfish" into the search box.
3. Choose your book to see a list of websites.

FACT SURFER